Microbiology for Complete Beginners

The Definitive Guide to
Understanding Microorganisms and
the Role of Microbes in Earth's
Ecosystem

Mina Mong
Copyright@2024

TABLE OF CONTENT

CHAPTER 1

Introduction to Microbiology

The field of microbiology focusses on the exploration of microorganisms—minute, frequently unseen life forms including bacteria, viruses, fungi, protozoa, and algae. These tiny entities, though minuscule, have an immense impact on the development of life on our planet. From the air we inhale to the nourishment we consume, and from our well-being to the balance of ecosystems, microorganisms play a crucial role in numerous biological functions. This introduction provides a comprehensive overview of microbiology, delves into its historical development, and emphasises the essential role of microorganisms across different aspects of life.

Definition of Microbiology

The field of microbiology, originating from the Greek terms mikros (small), bios (life), and logos (study), focusses on the scientific exploration of microorganisms. These minute life forms, typically only observable through a microscope, encompass a wide variety of organisms including bacteria, archaea, viruses, fungi, protozoa, and algae. Some microorganisms are unicellular, while others are multicellular, thriving in nearly every imaginable habitat on our planet, ranging from the depths of the oceans to the human body, and even in extreme conditions such as hot springs and polar ice caps.

The realm of microbiology is extensive. This field involves examining the structure, function, genetics, ecology, and evolution of microorganisms. This also explores the interactions of these organisms with their surroundings and hosts, encompassing humans, animals, plants, and various microbes. Researchers in the field explore the advantageous and detrimental functions of microorganisms, aiming to comprehend their influence on health, illness, industry, agriculture, and the ecosystem.

The study of microorganisms is essential in the scientific realm as it sheds light on core biological mechanisms. These processes are often more accessible to investigate in microbes, given their swift growth and straightforward genetic structures when contrasted with more complex organisms. It plays a crucial role in the realms of medicine, biotechnology, environmental science, and agriculture.

Chronicles of Microbial Science

The history of microbiology unfolds as a narrative of exploration and advancement, characterised by crucial milestones that have significantly transformed our comprehension of the natural environment.

Antonie van Leeuwenhoek (1632–1723), a Dutch tradesman and scientist, is frequently recognised as a pioneering figure in the study of microscopic life. With a meticulously crafted microscope, van Leeuwenhoek became the pioneer in observing and detailing microorganisms, which he called animalcules. His revolutionary findings, conducted in the late 1600s, encompassed the identification of microorganisms, single-celled organisms, and reproductive cells. While van Leeuwenhoek may not have grasped the full importance of his discoveries in that era, his contributions established a crucial groundwork for the advancement of microbiological research.

In the 19th century, Louis Pasteur (1822–1895), a French scientist, made several key contributions to microbiology. He is widely recognised for formulating the germ theory of disease, which suggested that numerous diseases are triggered by microorganisms. Pasteur's experiments with fermentation and spoilage revealed that microorganisms played a crucial role in these processes, effectively challenging the widely accepted theory of spontaneous generation, which posited that life could emerge spontaneously from non-living matter. He also pioneered the process of pasteurisation, a technique for eliminating harmful microbes in food and beverages, and formulated vaccines for diseases such as rabies and anthrax, which transformed public health.

Around the same time, Robert Koch (1843–1910), a German physician, made significant contributions to the field of medical microbiology. Koch is renowned for developing a series of criteria that are essential for determining the causal link between a microorganism and a disease. His investigations into ailments like tuberculosis, cholera, and anthrax contributed significantly to validating the germ theory of disease and paved the way for the creation of diagnostic methods that remain in use today. Koch's contributions significantly broadened the scope of microbiology and revolutionised medical practices.

Notable individuals in the development of microbiology encompass Joseph Lister, recognised for pioneering antiseptic surgery, and Alexander Fleming, celebrated for his discovery of the first antibiotic, penicillin. These advancements paved the way for deeper insights into infectious diseases, strategies for infection control, and the creation of crucial treatments that save lives.

Throughout the 20th and 21st centuries, the field has progressed significantly due to technological advancements and improvements in scientific methods. The elucidation of DNA's structure, the emergence of molecular biology methods, and the growth of genomics have broadened the horizons of microbiology. Researchers can now alter microorganisms at the genetic level, resulting in advancements in biotechnology, pharmaceuticals, agriculture, and environmental management.

The Significance of Microorganisms

Even though they are tiny, microorganisms play a crucial role in the ecosystem of our planet. They play an essential role in various processes that support ecosystems, human well-being, industry, and the natural world. Grasping the function of microbes is crucial for recognising their importance.

1. Microorganisms and Their Impact on Human Well-being

Microbes play a crucial role in maintaining human health. The human body hosts an immense number of microorganisms, which together form what is referred to as the human microbiome. The microbes that inhabit the gut, skin, mouth, and various other body locations carry out vital roles. Within the digestive system, microorganisms play a crucial role in breaking down food, contributing to vitamin production, and defending against harmful invaders. The equilibrium of these microorganisms is essential for sustaining health, and disturbances to the microbiome have been associated with conditions like obesity, diabetes, autoimmune diseases, and mental health issues.

Alongside helpful microbes, harmful microorganisms play a crucial role in causing infectious diseases. Microorganisms like Streptococcus pneumoniae are responsible for pneumonia, whereas viral agents such as the influenza virus result in the flu. Comprehending these microorganisms and their mechanisms of disease has facilitated the creation of therapies, immunisations, and health strategies that preserve countless lives each year. The study of microorganisms has been essential in combating diseases like tuberculosis, HIV/AIDS, and COVID-19.

2. Microorganisms in Their Ecosystems

Microorganisms play a crucial role in the functioning of Earth's ecosystems. These organisms play a crucial role in decomposing organic material, recycling essential nutrients, and facilitating biogeochemical processes, including the nitrogen and carbon cycles. For instance, certain bacteria have the ability to transform atmospheric nitrogen into usable forms for plants, highlighting the crucial role these microorganisms play in agriculture and the vitality of plant life.

In aquatic ecosystems, photosynthetic microbes such as cyanobacteria and algae generate oxygen and act as the cornerstone of the food chain. Marine microbes play a crucial role in breaking down pollutants, such as oil spills, via a process referred to as bioremediation.

Microorganisms contribute significantly to the regulation of climate systems. Some microorganisms generate greenhouse gases such as methane, whereas others have the ability to sequester carbon, contributing to the reduction of climate change impacts. Grasping the intricacies of microbial ecology is essential for effectively managing ecosystems, tackling environmental issues, and fostering sustainability.

3. Microorganisms in Industry

For millennia, humans have utilised microorganisms for a range of industrial purposes. Fermentation, facilitated by microorganisms such as bacteria and yeast, has been harnessed to create various products including bread, cheese, yoghurt, wine, and beer. Today, the field involves a diverse array of applications, such as the creation of antibiotics, vaccines, enzymes, biofuels, and biodegradable plastics.

The manipulation of microorganisms through genetic engineering in the field of biotechnology has resulted in the creation of innovative medicines, agricultural advancements, and technologies aimed at environmental improvement. For instance, the insulin utilised in diabetes management is now synthesised through the use of genetically modified bacteria. Microorganisms play a crucial role in waste treatment by breaking down organic matter found in sewage and industrial waste, thereby aiding in the creation of cleaner environments.

4. Microorganisms in Scientific Inquiry and Innovation

Microorganisms serve as essential instruments in the realm of scientific inquiry. Their straightforward nature, quick development, and clearly defined genetic frameworks render them perfect subjects for investigating essential life processes. A significant portion of our understanding of genetics, metabolism, and cellular processes has been derived from research involving organisms such as Escherichia coli and Saccharomyces cerevisiae.

The domain of synthetic biology, focused on the design and construction of novel biological systems, is significantly dependent on microorganisms. Researchers are manipulating microorganisms to generate sustainable energy, remediate environmental pollutants, and even synthesise medications. Progress in the study of microscopic life forms consistently expands the limits of potential in the fields of biotechnology and medicine.

The study of microbiology encompasses a broad and intriguing domain that delves into the invisible realm of microorganisms and their significant influence on life on our planet. Microbes play a crucial role in human health and disease, as well as in the essential functions of ecosystems and industry, yet their significance is frequently underestimated. Grasping the intricacies of microbiology enhances our appreciation for the complexity of life and prepares us to tackle significant challenges confronting humanity, including infectious diseases and environmental sustainability.

CHAPTER 2

The World of Microorganisms

Microorganisms, commonly known as microbes, represent the tiniest forms of life found on our planet. Even though they are minuscule, these small organisms are crucial to almost every facet of existence, from supporting ecosystems to impacting human well-being. This chapter explores the intriguing realm of microorganisms, detailing their definitions, various types, and their omnipresence in nature, including within the human body.

What Are Microorganisms?

Microorganisms are tiny life forms that cannot be observed without the aid of a microscope, as they are too minuscule for the naked eye to detect. These organisms encompass a diverse range, including bacteria, viruses, fungi, protozoa, and algae, each exhibiting distinct characteristics, life cycles, and functions within ecosystems. Comprehending the variety of microorganisms is essential to the field, as each type plays a crucial role in the vital processes that uphold life on our planet.

Understanding and Classifications of Microorganisms

1. Microorganisms:
Single-celled organisms known as bacteria are classified as prokaryotes and do not possess a defined nucleus. These organisms represent some of the earliest and most plentiful life forms on our planet, flourishing in a wide range of habitats. Microorganisms are essential participants in nutrient cycling, contributing to processes like nitrogen fixation, decomposition, and aiding in digestion within the human gastrointestinal system. Certain bacteria can be harmful, leading to illnesses such as tuberculosis, yet numerous types play a positive role, aiding in processes such as fermentation and biodegradation.

2. Pathogens:
Viruses are non-cellular entities, indicating that they lack the cellular structure found in bacteria, fungi, or protozoa. These entities consist of genetic material, either DNA or RNA, encased in a protein coat, and

occasionally enveloped by a lipid layer. Viruses lack the ability to replicate independently and require a host cell for reproduction. The parasitic characteristics of these entities are responsible for their role in causing a variety of diseases across humans, animals, and plants, including influenza, HIV/AIDS, and COVID-19. While they can be detrimental, viruses also contribute to gene transfer and the process of evolution.

3. Fungi:
Fungi are complex organisms characterised by the presence of a defined nucleus and various membrane-bound organelles. This category encompasses various forms of fungi, such as yeasts, moulds, and mushrooms. Fungi play an essential role as decomposers in ecosystems, facilitating the breakdown of dead organic matter and the recycling of nutrients. Certain fungi play a crucial role in the creation of food and medicinal products, exemplified by Penicillium, known for its production of the antibiotic penicillin. However, certain fungi can lead to infections in humans, such as Candida, which can cause thrush or athlete's foot.

4. Single-celled organisms:
Protozoa are unicellular, eukaryotic entities that display diverse modes of locomotion, utilising flagella, cilia, or pseudopodia. These organisms inhabit both water and land ecosystems, serving a crucial function in the food web by consuming bacteria and various microorganisms. Certain protozoa exhibit parasitic behaviour, leading to diseases such as malaria (Plasmodium), amoebic dysentery (Entamoeba histolytica), and sleeping sickness (Trypanosoma).

5. Photosynthetic organisms:
Algae are complex, photosynthetic organisms that vary from unicellular varieties such as Chlorella to larger, multicellular types like seaweed. In aquatic ecosystems, algae serve as essential primary producers, establishing the foundation of the food chain by harnessing sunlight to generate energy via photosynthesis. They contribute significantly to the global oxygen supply. Beyond their ecological significance, algae find applications in various industries for the production of biofuels, food additives, and pharmaceuticals.

Distinctions Between Prokaryotic and Eukaryotic Cells

Microorganisms can be divided into two primary groups according to their cellular structure: prokaryotic and eukaryotic.

1. Prokaryotic Cells:
Prokaryotes encompass bacteria and archaea, characterised by cells that do not possess a true nucleus or membrane-bound organelles. The genetic material, or DNA, resides in a nucleoid, which is an area within the cell that lacks a surrounding membrane. Prokaryotic cells generally exhibit a smaller and more straightforward organisation compared to eukaryotic cells, featuring components such as ribosomes, a cell membrane, and occasionally a cell wall. They multiply rapidly via binary fission and exhibit remarkable adaptability, thriving in a wide range of habitats.

2. Complex Cellular Structures:
Organisms classified as eukaryotes, which encompass fungi, protozoa, algae, plants, and animals, possess cells that are more intricate, featuring a genuine nucleus and various organelles like mitochondria,

endoplasmic reticulum, and Golgi apparatus. These structures perform distinct roles within the cell. Eukaryotic cells may exist as single-celled organisms or as part of multicellular structures, and their reproduction involves intricate processes such as mitosis or meiosis. Eukaryotes typically exhibit a larger size compared to prokaryotes and are often located in more specialised habitats.

Significance of Scale: Distinctions Between Microorganisms and Larger Organisms

Microorganisms are distinct from plants and animals mainly due to their size, structural characteristics, and ways of reproduction. Most microorganisms measure between 0.1 and 10 micrometres, significantly smaller than the typical size of plant and animal cells, which usually range from 10 to 100 micrometres. The diminutive dimensions of microorganisms result in an elevated surface area-to-volume ratio, facilitating effective nutrient uptake and accelerated metabolic activities. This is one explanation for the rapid reproduction of numerous microbes when conditions are favourable.

Due to their diminutive size and uncomplicated structure, microorganisms have evolved to thrive in harsh conditions, including scalding hot springs, deep-sea hydrothermal vents, and even in radioactive waste. Their basic structure enables them to save energy and flourish in conditions that would be inhospitable for more intricate life forms.

The Omnipresence of Microorganisms

Microorganisms inhabit every corner of our planet, thriving in both well-known and the most extreme habitats. Their widespread presence highlights their remarkable ability to adjust and their crucial function in the processes of life.

The Habitats of Microorganisms

1. Earth's substrate:
The soil is alive with a vast array of microbial organisms, housing billions of these tiny life forms in merely a handful of earth. Microorganisms such as bacteria, fungi, and protozoa present in the soil play a crucial role in decomposing organic matter, recycling essential nutrients, and

promoting plant growth through nitrogen fixation. Microorganisms in the soil are essential to the carbon cycle, as they break down deceased flora and fauna, releasing carbon back into the atmosphere and soil.

2. H2O:
Water-based ecosystems, such as seas, streams, ponds, and even ice formations, host a diverse array of microorganisms. In aquatic ecosystems, the foundational elements of food webs are composed of algae, cyanobacteria, and protozoa. In the oceans, tiny life forms generate oxygen via photosynthesis and play a crucial role in the global carbon cycle by taking in carbon dioxide. Marine microorganisms contribute significantly to climate regulation through the production of sulphur compounds, which have an impact on cloud formation.

3. Atmosphere:
Microorganisms, while less concentrated than in soil or water, are also present in the air. Wind can transport fungal spores, bacteria, and viruses, while plants and animals can also release them into the environment. Microorganisms in the air have the ability to traverse vast distances

and play a role in the dissemination of illnesses, yet they are also integral to essential processes such as nutrient cycling and the breakdown of organic materials.

4. Human Anatomy:
The human body hosts an immense number of microorganisms, which are referred to as the human microbiome. These microorganisms inhabit the skin, gut, mouth, and various other regions of the body, creating a complex and ever-changing ecosystem. The human microbiome is essential for digestion, immune system support, and safeguarding against harmful pathogens. For instance, microorganisms in the digestive system assist in the breakdown of intricate carbohydrates, synthesise essential vitamins such as B12 and K, and modulate immune reactions.

5. Harsh Habitats:
Microorganisms inhabit some of the most extreme environments on our planet, including hot springs, deep-sea hydrothermal vents, acidic lakes, and frozen polar regions. These remarkable organisms have adapted to thrive in environments that would be unlivable for the majority of

life forms. For example, certain bacteria flourish in elevated temperatures, whereas others are adapted to survive in saline conditions. Extremophiles serve as important subjects for scientific inquiry, aiding researchers in exploring the boundaries of life and providing valuable perspectives on the potential for life beyond our planet.

Exploring the Microbiome: An Overview

The concept of microbiome encompasses the diverse assembly of microorganisms residing within a specific environment, including the human body, soil, or oceanic ecosystems. The human microbiome has garnered considerable interest in recent years because of its substantial influence on health and disease.

1. The Human Microbiome:
The human body is home to an estimated 100 trillion microbes, which outnumber human cells by a ratio of about 10 to 1. These microorganisms carry out a variety of crucial roles. For example, microorganisms in the digestive tract assist in breaking down food, producing essential vitamins,

and modulating the immune response. They engage in a constant struggle with detrimental microorganisms for territory and nutrients, thereby thwarting infections.

2. Microbiome and Well-being:
The equilibrium of microorganisms within the human microbiome plays a vital role in sustaining health. Alterations to the microbiome, referred to as dysbiosis, have been associated with a range of health issues, such as obesity, diabetes, inflammatory bowel disease, and even mental health disorders. Investigations into the microbiome are revealing the ways in which these microbial communities affect not just physical health, but also mood, behaviour, and cognition.

Microorganisms exhibit remarkable diversity, adaptability, and omnipresence. These organisms thrive in a multitude of environments, ranging from the earth beneath our feet to the harsh conditions found at the depths of the ocean, and even within the human body. Exploring the realm of microorganisms reveals essential knowledge about the intricate relationships among living organisms on our planet, the dynamics of ecosystems, and the preservation of human well-being. As the field of microbiology advances, our understanding of the hidden microbial realm expands, uncovering the vital functions that these tiny organisms fulfil in maintaining life on Earth.

CHAPTER 3

Classification and Types of Microorganisms

Microorganisms are categorised into various groups according to their structural characteristics, life cycles, and ecological functions. The primary categories of microorganisms consist of bacteria, viruses, fungi, protozoa, and algae. Every one of these groups possesses specific traits and fulfils unique functions in the ecosystem and in relation to human well-being. This chapter will delve into the architecture, developmental trends, and roles of these microorganisms, while also examining their impacts on the ecosystem, human existence, and various industries.

Microorganisms: Anatomy, Morphology, and Developmental Trends

Microorganisms that consist of a single cell and lack a nucleus exhibit remarkable diversity and prevalence in various environments. These organisms inhabit nearly every ecosystem on our planet, ranging from the human body to the most profound depths of the ocean. Microorganisms are essential in ecosystems, as they decompose organic material, convert nitrogen into usable forms, and establish mutualistic interactions with flora and fauna.

Composition of Bacterial Cells

Bacteria possess a comparatively straightforward cellular architecture when contrasted with eukaryotic cells. The primary elements consist of:
1. Cell Membrane: A phospholipid bilayer that controls the movement of substances in and out of the cell.
2. Cell Wall: A sturdy framework composed of peptidoglycan that offers structural integrity and defence.
3. Cytoplasm: A viscous medium that houses the essential components required for metabolic processes.
4. Nucleoid: This is the area where bacterial DNA resides, and it is not surrounded by a membrane.
5. Ribosomes: Tiny entities that play a crucial role in the synthesis of proteins.
6. Flagella and Pili: These are specialised structures that facilitate movement and enable attachment to various surfaces.

Forms and Developmental Trends

Microorganisms can be categorised according to their morphological characteristics:
- Cocci: Spherical microorganisms (e.g., Staphylococcus aureus).
- Bacilli: These are bacteria characterised by their rod-like shape, such as Escherichia coli.
- Spirilla: Bacteria characterised by their spiral shape (e.g., Helicobacter pylori).

Bacteria exhibit diverse growth patterns:

- Individual microorganisms (e.g., E. coli).
- Chains (e.g., Streptococcus).
- Clusters (e.g., Staphylococcus).

Comparison of Bacteria Types: Gram-Positive and Gram-Negative

The categorisation of bacteria into two groups, Gram-positive and Gram-negative, hinges on the composition of their cell walls, as demonstrated through the Gramme staining technique:
- Gram-Positive Bacteria: Characterised by a robust peptidoglycan layer in their cell walls, these organisms retain the violet dye during the Gramme staining process, resulting in a purple appearance when observed under a microscope. Instances consist of Bacillus and Staphylococcus.
- Gram-Negative Bacteria: Characterised by a thinner peptidoglycan layer and an outer membrane that includes lipopolysaccharides. They fail to hold onto the violet dye, resulting in a pink or red appearance following the staining process. Instances encompass E. coli and Salmonella.

Functions in the Ecosystem

- Decomposition: Microorganisms dismantle deceased organic material, returning essential nutrients to ecosystems.
- Nitrogen Fixation: Specific microorganisms, like Rhizobium, transform atmospheric nitrogen into ammonia, making it available for plant growth.
- Symbiosis: Numerous microorganisms exist in mutually beneficial relationships with flora and

fauna. For example, the microorganisms in the human gut play a crucial role in breaking down food and synthesising vital vitamins.

Viruses: Composition and Development

Viruses possess a distinct characteristic as they do not qualify as entirely living entities. These are non-cellular organisms that can solely reproduce within a host cell, categorising them as mandatory parasites. Pathogens such as viruses play a crucial role in the emergence of various diseases, ranging from mild ailments like the common cold to more serious conditions including HIV and COVID-19.

Composition of Viral Particles

Viruses are composed of two primary elements:
1. Genetic Material: Either DNA or RNA, which contains the essential viral genes required for replication.
2. Protein Coat (Capsid): A protective layer composed of protein that surrounds the genetic material.

Certain viruses possess a lipid envelope encasing the capsid, which they obtain from the host cell membrane throughout the replication process.

Lifecycle: Lytic and Lysogenic Phases

Viruses exhibit two primary forms of lifecycles:
- Lytic Cycle: During this phase, the virus invades a host cell, commandeers its cellular machinery to

generate new viral particles, and eventually leads to the cell's rupture (lysis), allowing the newly formed viruses to spread and infect additional cells.
- Lysogenic Cycle: During this phase, the viral DNA becomes incorporated into the host cell's genetic material and stays inactive for a certain duration. It is replicated alongside the host cell's genetic material during cell division, until specific environmental factors prompt the virus to transition into the lytic cycle.

The Distinctions Between Viruses and Living Organisms

In contrast to bacteria, fungi, or protozoa, viruses are classified as non-living entities since they lack the ability to perform vital functions like metabolism or reproduction independently. They rely completely on a host cell for their replication, categorising them as obligate intracellular parasites.

Functions in Disease Development

Viruses are well-known for their role in inducing various diseases across humans, animals, and plants. For instance:
- Influenza Virus: Responsible for the flu, a highly transmissible respiratory ailment.
- Human Immunodeficiency Virus (HIV): Causes acquired immunodeficiency syndrome (AIDS) by specifically attacking the immune system.
- SARS-CoV-2: The pathogen that led to the global COVID-19 outbreak.

Fungi: Yeasts, Moulds, and Mushrooms

Fungi are complex organisms that can be found in both unicellular forms, such as yeasts, and multicellular forms, including moulds and mushrooms. They serve crucial functions within ecosystems, especially in their capacity as decomposers. Fungi play a crucial role in various industries and medicine, particularly in the synthesis of antibiotics and the process of fermentation.

Varieties of Fungi

1. Yeasts: Unicellular organisms that propagate through the process of budding. Yeasts play a crucial role in the fermentation process, facilitating the production of alcohol and bread. An illustrative case is Saccharomyces cerevisiae, which finds application in both brewing and baking processes.
2. Moulds: Complex organisms that develop as networks of thread-like structures known as hyphae. Moulds such as Aspergillus and Penicillium play a crucial role in the creation of antibiotics (penicillin) and various food items (cheese).
3. Fungi: The fruiting bodies of certain fungi, commonly known as mushrooms, are multicellular organisms that can either be edible or toxic. They are essential in the decomposition of organic material within ecosystems.

Significance in Natural Environments and Commercial Sectors

- Decomposition: Fungi break down deceased flora and fauna, recycling nutrients back into the soil and supporting ecological equilibrium.
- Fermentation: Microorganisms play a crucial role in the creation of alcoholic drinks and the leavening of bread. Fungi play a crucial role in the creation of soy sauce, cheese, and various other fermented foods.
- Antibiotics: Certain organisms generate natural antibiotics, like penicillin, which is utilised in the treatment of bacterial infections.

Protozoa and Algae: Unicellular Eukaryotic Life Forms

Single-celled eukaryotic organisms such as protozoa and algae play a variety of important roles within ecosystems. Protozoa frequently have connections to illness, whereas algae play an essential role in oxygen generation and the process of photosynthesis.

Single-celled organisms

Protozoa are dynamic, unicellular entities that may exist independently or as parasites. These organisms utilise flagella, cilia, or pseudopodia to facilitate their movement. Protozoa are crucial components of aquatic ecosystems, acting as predators of bacteria and other microorganisms, while some also contribute to significant diseases.

- As Parasites: Some protozoa exhibit parasitic behaviour, leading to diseases in both humans and animals. Illustrations encompass:
- Plasmodium: Responsible for malaria, it is spread via the bite of a mosquito that carries the infection.
- Entamoeba histolytica: Responsible for amoebic dysentery, resulting in intense diarrhoea.
- Trypanosoma: Responsible for sleeping sickness, spread by the tsetse fly.

Photosynthetic organisms

Photosynthetic organisms known as algae inhabit aquatic environments and contribute a substantial amount of the Earth's oxygen supply. These organisms vary from unicellular varieties such as Chlorella to extensive multicellular types like seaweed. Algae play an essential role in maintaining the vitality of aquatic ecosystems, serving as the foundation of the food chain and generating oxygen via photosynthesis.

- Function in Energy Conversion and Oxygen Release: Algae harness sunlight to produce energy via photosynthesis, emitting oxygen as a byproduct. This process is essential for sustaining oxygen levels in the atmosphere and fostering life on our planet.

Microorganisms, such as bacteria, viruses, fungi, protozoa, and algae, play a crucial role in the ecosystem of our planet. The categorisation into various groups enables researchers to examine their distinct traits and comprehend their roles in ecosystems, industry, and human health. Through the analysis of the structure, functions, and effects of these tiny organisms, we develop a greater understanding of the minuscule realm that affects everything from the air we inhale to the sustenance we consume.

CHAPTER 4

The Microbial Ecosystem

Microorganisms, despite their invisibility to the naked eye, are crucial for the proper functioning of ecosystems on our planet. These tiny organisms are not alone; they thrive within intricate and varied communities, engaging with one another and their surroundings in countless interactions. From enabling nutrient cycling to thriving in extreme environments, microorganisms are essential to the existence of life on our planet. This chapter delves into the fascinating world of microbial communities, highlighting their essential functions in nutrient cycles and their remarkable capacity to flourish in extreme environments.

Microbial Ecosystems

Microbial communities consist of various microorganisms that live together and engage with one another in a common environment. These communities exist in nearly every environment on Earth, ranging from soil and water to the human body. In these communities, microorganisms frequently collaborate or vie for dominance, influenced by environmental factors and the availability of resources.

Exploring the Interactions of Microbes with One Another and Their Surroundings

Microorganisms engage with one another through a range of mechanisms, such as:

- Symbiosis: Certain microorganisms engage in mutually advantageous relationships, either among themselves or with larger life forms. For instance, certain bacteria inhabit plant roots, transforming nitrogen into a usable form for plants, while the plants supply these microorganisms with essential nutrients.

- Competition: In certain ecosystems, the availability of resources is limited, prompting microorganisms to vie for nutrients, habitat, and energy sources. Microorganisms can produce substances, including antibiotics, to suppress the proliferation of competing species.

- Quorum Sensing: Numerous microorganisms engage in communication via chemical signals in a phenomenon referred to as quorum sensing. This allows for the coordination of behaviour, such as the formation of biofilms, influenced by population density.

Biofilms: Development and Significance in Health and Industry

One of the most fascinating instances of microbial collaboration is the development of biofilms. Microbial communities known as biofilms attach to surfaces and generate a protective matrix made up of extracellular polymeric substances (EPS), which include proteins, sugars, and nucleic acids.

- Formation: Biofilms initiate when free-floating bacteria adhere to a surface and secrete extracellular polymeric substances, forming a cohesive environment that facilitates the inclusion of additional microorganisms into the community. Once established, these microbial communities exhibit remarkable resilience against external challenges, including antibiotics and disinfectants.

- Significance in Health: Biofilms play a crucial role in human health. A prominent illustration is dental plaque, a biofilm that develops on teeth and can result in cavities and gum disease. Biofilms can develop on medical devices, including catheters and prosthetic joints, resulting in persistent infections that are challenging to manage because of their resistance to antibiotics.

- Significance in Industry: Biofilms are involved in numerous industrial processes, influencing outcomes in both beneficial and detrimental ways. In the process of wastewater treatment, microbial communities are employed to decompose organic substances. Nonetheless, they can lead to complications in sectors such as food processing and energy by obstructing pipelines or tainting surfaces.

The Function of Microorganisms in Nutrient Recycling

Microbes play a crucial role in global nutrient cycles, converting essential elements such as nitrogen, carbon, and sulphur into forms accessible to plants, animals, and various other organisms. The intricate interactions of these microorganisms play a crucial role in sustaining nutrient equilibrium within ecosystems, ultimately supporting the overall vitality of our planet.

The Role of Microorganisms in the Transformation of Key Elements: Nitrogen, Carbon, and Sulphur Cycles

- Nitrogen Cycle: Microorganisms are essential in the nitrogen cycle, a process that transforms nitrogen into different chemical forms that are accessible to living organisms. For instance, certain bacteria are capable of transforming atmospheric nitrogen (N_2) into ammonia (NH_3), making it accessible for plant absorption. During nitrification, certain microorganisms transform ammonia into nitrates (NO_3^-), a form that is accessible for plant uptake. Ultimately, in the process of denitrification, specific bacteria transform nitrates into atmospheric nitrogen, thereby concluding the cycle.

- Carbon Cycle: Microorganisms play a crucial role in the carbon cycle, which governs the movement of carbon within ecosystems. During decomposition, microorganisms such as bacteria and fungi play a crucial role in breaking down dead organic matter, which results in the release of carbon dioxide (CO_2) into the atmosphere. In contrast, photosynthetic organisms such as algae and cyanobacteria absorb CO_2 from the atmosphere and utilise it to generate energy via the process of photosynthesis.

- Sulphur Cycle: Sulphur plays a crucial role in the ecosystem, and microorganisms are key players in the processes that drive the sulphur cycle. Some microorganisms, like Desulfovibrio, have the ability to transform sulphate (SO_4^{2-}) into hydrogen sulphide (H_2S), a phenomenon referred to as sulphate reduction. Some microorganisms convert hydrogen sulphide back into sulphate, thereby finalising the sulphur cycle. The conversion of sulphur compounds plays a crucial role in sustaining the equilibrium of sulphur within ecosystems, thereby fostering the vitality of both plant and microbial communities.

Microorganisms in Soil Decomposition

Beyond their involvement in nutrient cycling, microorganisms serve as decomposers, dismantling deceased plants, animals, and organic matter within the soil. This process enhances the soil by adding vital nutrients, including nitrogen, phosphorus, and potassium, which are crucial for the growth of plants. Microorganisms involved in decomposition, including bacteria and fungi, are essential for the creation of humus, an organic matter that enhances soil fertility and its ability to retain water.

The breakdown of organic matter by microorganisms plays a crucial role in maintaining soil vitality and facilitating the capture of carbon. Through the decomposition of plant matter and the integration of carbon into the soil, microorganisms play a crucial role in lowering atmospheric carbon dioxide levels, thereby helping to alleviate the impacts of climate change.

Microorganisms in Harsh Conditions

While many microbes flourish in moderate environments, certain microorganisms, referred to as extremophiles, have adapted to survive in extreme conditions, including high temperatures, high acidity, high salinity, or even in oxygen-deprived settings. These extremophiles provide important understanding of life's adaptability and have significant applications in biotechnology and space exploration.

Extremophiles: Microorganisms Thriving in Harsh Environments

- Thermophiles: These microorganisms, which flourish in high temperatures exceeding 45°C (113°F), inhabit environments like hot springs, hydrothermal vents, and various industrial processes. Thermus aquaticus, a heat-loving bacterium, is significant for its contribution to the polymerase chain reaction (PCR) method utilised in the study of biological processes at the molecular level.

- Halophiles: These fascinating microorganisms thrive in saline environments, demonstrating remarkable adaptations that allow them to flourish in areas with elevated salt concentrations, like salt lakes or salt mines. Halobacterium exemplifies an organism that flourishes in high-salt environments,

showcasing the remarkable adaptability of certain life forms.
- Organisms thriving in acidic and alkaline environments: Organisms that flourish in extremely acidic conditions can be found in environments like sulphuric acid springs, whereas those that prefer basic conditions are often located in soda lakes. These microorganisms possess unique adaptations that enable them to regulate their internal pH and thrive in extreme environments.

- Psychrophiles: These microorganisms thrive in frigid conditions, enduring temperatures that dip below 0°C (32°F). They inhabit extreme environments such as polar ice caps, glaciers, and the depths of ocean waters.

The Advantages of Researching Extremophiles for Advancements in Biotechnology and Space Exploration

- Biotechnology: Organisms that thrive in extreme environments generate enzymes and proteins that operate under harsh conditions, rendering them essential for industrial and biotechnological uses. For instance, the enzyme Taq polymerase, sourced from the heat-loving bacterium Thermus aquaticus, plays a crucial role in the PCR method utilised in genetic studies. Other extremophiles generate enzymes that are beneficial in sectors like food processing, bioremediation, and biofuel production.

- Space Exploration: Organisms that thrive in extreme conditions also have implications for the search for extraterrestrial life. Investigating the survival mechanisms of microorganisms in Earth's extreme environments allows for a deeper comprehension of the possibilities for life in the harsh conditions found on other planets or moons, including Mars and Europa. Investigating extremophiles provides valuable insights for research in astrobiology and aids in creating technologies aimed at identifying life outside our planet.

Microbes play a crucial role in Earth's ecosystems, creating communities that engage with their surroundings and one another. Their functions in nutrient cycling and decomposition are vital for sustaining life, and their capacity to thrive in extreme environments presents exciting opportunities in biotechnology and space exploration. Exploring the microbial ecosystem uncovers the complex interplay of life on our planet and the immense possibilities that microorganisms offer for future scientific progress.

CHAPTER 5

Microorganisms and Human Health

Microorganisms are crucial to human health, serving as both helpful allies and possible dangers. The extensive community of microorganisms residing on and within our bodies, referred to as the human microbiome, plays a crucial role in functions such as digestion, immune response, and even mental health. Conversely, harmful microorganisms—bacteria, viruses, fungi, and protozoa that lead to diseases—represent considerable threats to human health, requiring a strong immune reaction and the advancement of medical solutions such as antibiotics and vaccines. This chapter delves into the intricate interplay between tiny organisms and human well-being, emphasising the human microbiome, harmful microbes, and the significance of vaccines in disease prevention.

The Human Microbiome

The human microbiome encompasses the varied assembly of microorganisms, such as bacteria, viruses, fungi, and archaea, that reside in different regions of the human body. These microorganisms inhabit the gut, skin, mouth, respiratory tract, and even the reproductive system. Rather than posing a threat, the majority of these microorganisms are advantageous and are essential for sustaining our well-being.

Microorganisms in Various Body Systems: Gut, Skin, Mouth, and Beyond

- Gut Microbiome: The human digestive system hosts an immense variety of microorganisms that play a crucial role in breaking down food and facilitating nutrient uptake. Essential microbial species, including Bacteroides and Firmicutes, play a crucial role in the breakdown of complex carbohydrates, fibres, and other indigestible substances, converting them into short-chain fatty acids (SCFAs) such as butyrate, which support the gut lining and modulate inflammation. The gut microbiome plays a crucial role in synthesising vital vitamins, including B vitamins and vitamin K.

- Skin Microbiome: The skin acts as a vital protective barrier and is home to a diverse community of microbes, including bacteria such as Staphylococcus epidermidis and Propionibacterium acnes. These microorganisms safeguard the skin by outcompeting detrimental pathogens, regulating skin pH, and enhancing immune responses.

- Oral Microbiome: The mouth hosts a diverse array of microorganisms that play a crucial role in the initial phases of digestion and help inhibit the proliferation of detrimental microbes. Certain species contribute to the formation of biofilms on teeth, known as dental plaque, while additional bacteria present in saliva secrete enzymes that facilitate the breakdown of food particles.
- Respiratory and Urogenital Microbiomes: The microbiome of the respiratory tract plays a crucial

role in defending against respiratory infections. In a similar vein, the urogenital system, especially in females, hosts advantageous microorganisms such as Lactobacillus, which plays a crucial role in preserving the acidic pH of the vagina and inhibiting the proliferation of detrimental pathogens.

Their Role in Digestion, Immune Function, and General Well-Being

- Digestion: The gut microbiome plays a crucial role in the processes of digestion and the absorption of nutrients. Microorganisms decompose intricate fibres into short-chain fatty acids, which play a crucial role in managing metabolism and energy equilibrium. A well-balanced gut microbiome is associated with reduced risks of digestive issues such as irritable bowel syndrome (IBS) and inflammatory bowel disease (IBD).

- Immunity: Microorganisms are crucial for the development and functioning of the immune system. The immune system is in a continuous dialogue with the microbiome, developing the ability to distinguish between detrimental and benign microbes. Helpful microorganisms assist in educating immune cells and sustaining a balanced immune reaction, thereby lowering the chances of allergies, autoimmune disorders, and persistent inflammation.

- Overall Health: A well-maintained microbiome contributes to comprehensive wellness, encompassing mental health as well. The gut-brain axis describes the intricate communication system linking the gut microbiome and the brain, impacting mood, behaviour, and stress responses. Disruptions in the gut microbiome, often referred to as dysbiosis, have been associated with various mental health issues, such as depression and anxiety.

Exploring the Connection Between the Microbiome and Conditions Such as Obesity, Diabetes, and Autoimmune Disorders

Alterations to the microbiome can play a role in a range of health issues:

- Obesity: Studies indicate that those with obesity frequently exhibit a distinct microbial makeup in contrast to their lean counterparts. An imbalance between Firmicutes and Bacteroidetes has been linked to weight gain. These microorganisms play a crucial role in the efficiency of energy extraction from food, which may have implications for fat accumulation and metabolic conditions.

- Diabetes: An imbalance in the gut microbiome, known as dysbiosis, has been associated with type 2 diabetes. Some microorganisms have the potential to influence insulin sensitivity, inflammation, and glucose metabolism, playing a role in the onset of insulin resistance and elevated blood sugar levels.

- Immune System Malfunctions: An imbalanced microbiome can trigger an exaggerated immune response, potentially playing a role in autoimmune conditions such as rheumatoid arthritis, multiple sclerosis, and lupus. The decline in microbial diversity impairs the immune system's capacity to differentiate between the body's own cells and harmful invaders, which could lead to an autoimmune response.

Disease-Causing Microorganisms

While many microbes provide advantages, some microorganisms are harmful, indicating their potential to induce diseases in humans. These microorganisms encompass bacteria, viruses, fungi, and protozoa, with their capacity to infect and affect humans differing significantly based on the species and the host's immune reaction.

The Mechanisms by Which Various Microorganisms Induce Illness

- Bacteria: Harmful bacteria are responsible for various illnesses. For instance, Streptococcus pneumoniae is responsible for pneumonia, whereas certain strains of Escherichia coli (E. coli) can result in food poisoning and urinary tract infections. Microorganisms can lead to illness through the secretion of harmful substances, the destruction of host cells, or the activation of an overzealous immune reaction.

- Viruses: These entities lack the characteristics of life and rely on host cells for their replication process. Viral infections are responsible for diseases such as influenza, HIV/AIDS, and COVID-19. Viruses commandeer host cells, utilising them for replication and dissemination, frequently resulting in the demise of the host cell during the process. Certain viruses, such as the human immunodeficiency virus (HIV), compromise the immune system, rendering the host vulnerable to additional infections.

- Fungi: Although fungi contribute positively to ecosystems, some species can pose health risks to humans. For instance, Candida is known to lead to yeast infections, whereas Aspergillus can result in respiratory infections in those with weakened immune systems. Fungi induce disease through tissue invasion or by producing harmful metabolites.

The Immune System's Protective Strategies Against Microbial Infections

The immune system has developed intricate mechanisms to safeguard the body against microbial threats. This can be categorised into two primary elements:
- Innate Immunity: This serves as the initial barrier the body employs to combat infections. Innate immune responses encompass physical barriers such as skin and mucous membranes, chemical barriers like stomach acid and enzymes, as well as immune cells including macrophages and

neutrophils that identify and eliminate invading microbes.

- Adaptive Immunity: When a pathogen evades the innate immune system, the adaptive immune response is activated. This entails distinct immune cells, such as T cells and B cells, which possess the ability to identify particular pathogens and retain a memory of them for more rapid responses in subsequent encounters. B cells generate antibodies, which are proteins that attach to microbes and render them harmless.

Overview of Antimicrobial Therapies (Antibiotics, Antivirals, Antifungals)

- Antibiotics: These are substances that are designed to combat infections caused by bacteria. These agents function by suppressing the proliferation of bacteria or directly eliminating them. Nonetheless, the excessive and improper use of antibiotics has resulted in antibiotic resistance, complicating the treatment of bacterial infections.

- Antivirals: These medications are employed to combat viral infections. In contrast to antibiotics, these agents do not eliminate viruses; instead, they inhibit their replication process. For instance, antiviral medications such as oseltamivir (Tamiflu) are utilised in the treatment of influenza, whereas antiretroviral therapies (ARVs) are employed to control HIV infections.

- Antifungals: These medications are employed to combat infections caused by fungi. These agents function by interfering with the cellular membrane of the fungi, leading to the demise of the cells. Typical treatments for fungal infections encompass topical applications for skin issues and systemic medications for more severe cases.

Vaccines and Immunisation

Vaccines represent a highly effective strategy for the prevention of infectious diseases. These function by activating the immune system to generate a defence that safeguards against subsequent infections, typically without inducing disease.

The Mechanisms Behind Vaccines in Combating Viral and Bacterial Infections

Vaccines consist of weakened or inactivated versions of a pathogen, or components like proteins, that stimulate an immune response while preventing the onset of disease. This response involves the generation of antibodies and the stimulation of memory cells, which retain the memory of the pathogen. When an individual who has received vaccination encounters the actual pathogen later on, their immune system is primed to respond swiftly, thereby averting the onset of illness.

The Influence of Vaccines on Community Wellness

The development of vaccines has significantly influenced global health, leading to a marked decrease in the occurrence of diseases like smallpox, polio, measles, and tetanus. Indeed, smallpox has been completely eliminated due to a worldwide vaccination initiative. Extensive vaccination initiatives persist in preserving countless lives annually, alleviating the impact of infectious illnesses and safeguarding at-risk groups.

Nonetheless, reluctance towards vaccines and the spread of false information have created obstacles for public health initiatives, highlighting the critical need for education and confidence in scientific knowledge to achieve broad immunisation and prevent disease.

Microorganisms play dual roles as both partners and challengers in the pursuit of human health. The interplay between the beneficial microbes residing in our microbiome and the pathogenic bacteria, viruses, and fungi responsible for diseases reveals a complex relationship with human health. Vaccines, antimicrobial treatments, and a deeper comprehension of the human microbiome are crucial instruments in preserving health and fighting disease.

CHAPTER 6

Microbiology in Everyday Life

Microorganisms are essential components of daily life, fulfilling vital functions in food production, industrial activities, and agricultural practices. Although numerous individuals link microorganisms with disease, their beneficial roles frequently remain overlooked. From fermenting food and producing lifesaving antibiotics to enhancing soil health and cleaning up environmental pollutants, these tiny organisms are essential for sustaining human life and the ecosystem. This chapter delves into the real-world uses of microbiology in food production, industry, and agriculture, emphasising the influence of microorganisms on our everyday lives.

Microbial Aspects of Food

Microorganisms have been utilised for ages in the processes of food creation, preservation, and ensuring safety. They serve a dual purpose in this field—facilitating the production of appealing food items through fermentation, while also playing a part in food spoilage and the occurrence of foodborne illnesses. Grasping the management and application of microbes in food is crucial for the food sector and the well-being of the public.

Microorganisms in Culinary Processes: Fermentation (Bread, Cheese, Yoghurt, Beer)

Fermentation represents a time-honoured practice within the realm of microbiological processes, harnessing the metabolic functions of microorganisms to transform sugars and various substrates into acids, gases, or alcohol. This process plays a crucial role in the development of a variety of food products:

- Bread: The creation of bread relies on Saccharomyces cerevisiae, commonly known as baker's yeast, which ferments sugars in the dough to generate carbon dioxide. This gas facilitates the expansion of the dough, leading to the light and airy quality characteristic of bread.

- Cheese and Yoghurt: Lactic acid bacteria like Lactobacillus and Streptococcus play a crucial role in the fermentation of dairy products. In the process of cheese production, certain bacteria transform lactose into lactic acid, facilitating the curdling of milk and ultimately resulting in the formation of cheese. In yoghurt production, lactic acid bacteria play a crucial role by fermenting the lactose found in milk. This process not only thickens the milk but also imparts the distinctive tangy flavour that yoghurt is known for.

- Beer and Wine: The process of alcoholic fermentation is facilitated by yeasts, especially Saccharomyces cerevisiae, which transform sugars derived from grains (in beer) or grapes (in wine) into alcohol and carbon dioxide. This fermentation process produces the effervescence in beer and contributes to the alcoholic content in wine.

Fermented foods are appreciated not just for their unique tastes but also for the positive effects they can have on health. Beneficial microorganisms present in fermented foods such as yoghurt play a crucial role in promoting gut health by enhancing digestion and maintaining a balanced gut microbiome.

The Function of Microorganisms in Food Decomposition and Illnesses Related to Food Consumption

While fermentation highlights the advantageous role of certain microbes, other microorganisms play a significant part in food spoilage and the occurrence of foodborne illnesses. Decomposition happens when microorganisms such as bacteria, yeasts, or moulds degrade food, resulting in unpleasant flavours, odours, textures, and potential health risks.

- Food Spoilage: Certain bacteria such as Pseudomonas and Clostridium flourish in decomposing food, leading to unpleasant odours, slimy textures, and colour changes. Fungi like Penicillium and Aspergillus can proliferate on food, leading to noticeable deterioration. These microorganisms break down food quality, rendering it unsafe or unappetising for consumption.

- Foodborne Illnesses: Harmful microorganisms, including Salmonella, Escherichia coli (E. coli), and Listeria, have the potential to taint food and lead to illnesses upon consumption. These microorganisms frequently arise from inadequate handling, storage, or contamination of unprocessed materials. Foodborne illnesses can result in a spectrum of symptoms, from mild digestive issues to severe, potentially life-threatening conditions.

Methods of Preservation: Pasteurisation, Refrigeration, and Canning

To prevent spoilage and foodborne illnesses, various methods are employed to inhibit microbial growth or eliminate harmful pathogens:

- Pasteurisation: This method, named after Louis Pasteur, entails heating food (especially liquids such as milk) to a precise temperature for a designated duration to eliminate harmful bacteria while preserving the food's flavour and nutritional integrity.

- Refrigeration: Low temperatures inhibit the growth of microorganisms, extending the shelf life of perishable items. Refrigeration plays a crucial role in preserving perishable foods like dairy products, meats, and fruits, as it significantly extends their shelf life by preventing the proliferation of microorganisms that lead to spoilage.

- Canning: This technique entails enclosing food in sealed containers and applying heat to eliminate any existing bacteria, yeasts, or moulds. The process of canning establishes a controlled environment within the container, enabling food preservation over long durations.

Applied Microbiology

Microorganisms play a crucial role in various industrial processes, contributing to the production of valuable products, waste treatment, and the remediation of environmental pollutants. Industrial microbiology emphasises the utilisation of microbial processes for extensive production and the promotion of environmental sustainability.

Application of Microorganisms in Biotechnological Processes: Synthesis of Antimicrobials, Catalysts, and Renewable Energy Sources

Microorganisms play a crucial role in biotechnology, contributing to the development of pharmaceuticals, enzymes, and sustainable energy solutions.

- Antibiotics: Organisms such as Penicillium (a mould) and Streptomyces (a bacteria) play a vital role in the production of antibiotics. Penicillium played a crucial role in the discovery of penicillin, which became one of the earliest and most extensively utilised antibiotics. These microorganisms inherently generate antibiotics to suppress the proliferation of rival bacteria, and researchers have utilised this capability for medical applications in humans.

- Enzymes: Microorganisms such as bacteria and fungi generate enzymes that find utility in a range of industrial processes, encompassing food processing, textiles, and the production of biofuels. For instance, enzymes such as amylase and protease are incorporated in detergents to decompose starches and proteins.

- Biofuels: Specific microorganisms, including Clostridium species, have the ability to transform organic waste into biofuels such as ethanol and methane. These microorganisms break down plant material or organic waste, creating a sustainable energy source that can aid in decreasing our reliance on fossil fuels.

Waste Treatment and Bioremediation: The Role of Microorganisms in Pollution Cleanup and Wastewater Management

Microorganisms are essential in the processes of waste management and environmental cleanup, as they effectively break down or eliminate toxic substances.

- Wastewater Treatment: Microorganisms play a crucial role in wastewater treatment facilities by decomposing organic materials and neutralising toxic substances found in sewage. The activity of microorganisms plays a crucial role in eliminating pollutants, resulting in cleaner water that is safe for environmental release.

- Bioremediation: Microorganisms are utilised to remediate environmental contaminants, including oil spills, heavy metals, and hazardous chemicals. Some microorganisms possess the capability to break down or alter these contaminants into less toxic forms, rendering them essential agents for ecological rehabilitation.

Soil and Plant Microbiology

In agriculture, tiny life forms play a crucial role in enhancing soil vitality, supporting plant development, and managing pests. Agricultural microbiology examines the relationships between microorganisms and plants, with the goal of improving food yield and promoting sustainable practices.

Microorganisms in Soil Vitality and Agricultural Yield

Soil is rich with microbial organisms, such as bacteria, fungi, and protozoa, which are essential for enhancing soil fertility and boosting crop productivity. These microorganisms break down organic material, liberating vital nutrients into the soil and facilitating their absorption by plants.

- Decomposers: Microorganisms such as bacteria and fungi in the soil play a crucial role in breaking down dead plant material and organic waste, thereby releasing nitrogen, phosphorus, and other vital nutrients necessary for the growth of plants. In the absence of these essential organisms, soil would rapidly lose its nutrient content, hindering the growth of crops.

The Importance of Helpful Microorganisms in Enhancing Plant Development (Mycorrhizae, Rhizobia)

Some microorganisms establish mutualistic associations with plants, enhancing their ability to absorb nutrients effectively.

- Mycorrhizae: These fungi engage in symbiotic partnerships with the roots of plants, enhancing nutrient uptake and overall health. Mycorrhizal fungi enhance the surface area of plant roots, enabling greater absorption of water and nutrients, especially phosphorus, from the soil. The plant supplies the fungi with carbohydrates in return.

- Rhizobia: These bacteria capable of fixing nitrogen establish symbiotic associations with leguminous plants such as beans and peas. They transform atmospheric nitrogen into a usable form for plants, enhancing the soil with this vital nutrient. This method minimises reliance on artificial nitrogen fertilisers, fostering environmentally friendly farming practices.

Natural Pest Management Through Microbial Solutions
Microorganisms serve as natural agents for managing agricultural pests, offering an environmentally sustainable option compared to chemical pesticides.

- Bacillus thuringiensis (Bt) is a bacterium frequently utilised as a natural pest control agent. It generates substances that are detrimental to insect pests while remaining harmless to humans and other animals. Upon consumption by pests, Bt interferes with their digestive processes, leading to their demise.

- Fungal Biopesticides: Specific fungi, such as Beauveria bassiana, are utilised for managing insect populations. These fungi target and eliminate pests such as aphids, caterpillars, and whiteflies, assisting farmers in safeguarding their crops without the need for synthetic chemicals.

Microorganisms are intricately connected to our daily existence, shaping food production, industrial activities, and agricultural practices. Their diverse functions—from fermenting foods and generating antibiotics to improving soil health and managing pests—highlight the significance of this field in fostering a sustainable and thriving environment.

CHAPTER 7

Microorganisms and the Environment

Microorganisms serve as the invisible builders of our planet's ecosystems, crucially influencing nutrient cycling, climate regulation, and the well-being of both land and water environments. From the depths of the oceans to the expansive stretches of forests, microorganisms play a crucial role in maintaining the stability and resilience of ecosystems. This chapter explores the intricate realm of environmental microbiology, focussing on the role of microbes in climate change, the mitigation of environmental pollution, and the essential operations of marine ecosystems.

Ecological Microbiology

Environmental microbiology involves examining the role of microorganisms in various natural settings, including soil, water, and the atmosphere. Microorganisms play a crucial role in ecosystem dynamics, ensuring nutrient balance, decomposing organic materials, and sustaining various life forms throughout the food web.

The Essential Function of Microorganisms in Ecosystems, from Oceans to Forests

Microorganisms like bacteria, fungi, archaea, and algae play essential roles in sustaining the health and functionality of ecosystems. These minute entities serve as the cornerstone of cycles that transport vital elements such as carbon, nitrogen, and sulphur throughout the ecosystem.

- In forests, microorganisms play a crucial role in breaking down organic matter (such as fallen leaves and dead animals), transforming it into simpler compounds that plants can utilise as nutrients. This process enhances soil quality and fosters robust plant development, highlighting the essential role of microbes in forest ecosystems.

- In the vast expanse of the seas, tiny organisms such as phytoplankton, which are photosynthetic algae, establish the foundation of the marine food web, supplying energy to higher trophic levels including zooplankton, fish, and marine mammals. In the absence of microbial activity, the very foundation of marine ecosystems would disintegrate.

- In wetlands and peat bogs, specific microorganisms contribute to the regulation of water quality and nutrient levels, while also participating in carbon sequestration—extracting carbon dioxide from the atmosphere and storing it in plant material and soils.

The Importance of Microbial Diversity for Ecosystem Stability and Resilience

The variety of microbial species in an environment is crucial for ecosystem stability. Ecosystems characterised by significant microbial diversity exhibit greater resilience to disruptions such as climate change, pollution, and habitat destruction.

- Microorganisms carry out a range of ecosystem roles, and an increase in species diversity enhances the breadth of these functions. For instance, certain microorganisms are capable of nitrogen fixation, whereas others play a crucial role in the decomposition of organic materials. When a species is lost from an ecosystem, another may step in to fulfil its role, helping to maintain the overall functionality of the ecosystem.

- Varied microbial populations enhance nutrient cycling efficiency and provide a defence against pathogen invasions. In agricultural soils, a diverse array of microorganisms can diminish the reliance on chemical fertilisers and pesticides, thereby enhancing the sustainability of farming practices.

Microorganisms and Climate Dynamics

Microorganisms play a crucial role in both the enhancement and reduction of climate change effects. They engage in the intricate processes of the carbon cycle and the generation of greenhouse gases, while also presenting promising avenues for mitigating environmental pollution and achieving climate stability.

The Role of Microorganisms in Carbon Cycling and Greenhouse Gas Emissions

Microbes play a crucial role in the carbon cycle, which encompasses the transfer of carbon among the atmosphere, oceans, soil, and living organisms. Carbon is vital for life, yet its uneven distribution—particularly the rise in carbon dioxide (CO_2) concentrations in the atmosphere—fuels global warming.

- Photosynthetic microorganisms such as cyanobacteria and phytoplankton play a crucial role in sequestering CO_2 from the atmosphere via photosynthesis, transforming it into biomass. These life forms establish the base of the food web and capture significant quantities of carbon in both oceanic and terrestrial environments.

- Decomposer microbes play a crucial role in breaking down organic matter (like deceased plants and animals) and releasing CO_2 back into the atmosphere through the process of respiration. In certain ecosystems, like peatlands and wetlands, the activity of decomposers is hindered by waterlogged conditions, leading to the buildup of organic material and the storage of carbon.

- Methanogens, a category of archaea, generate methane (CH_4), a significant greenhouse gas, in oxygen-free settings such as wetlands, rice fields, and the digestive tracts of ruminant creatures (e.g., cows). Methane is about 25 times more efficient at capturing heat in the atmosphere compared to CO_2, highlighting the importance of microbial methane production as a key factor in climate change.

The Function of Microorganisms in Alleviating Environmental Contamination (e.g., Oil Spills)

While certain microbes play a role in greenhouse gas production, others assist in reducing pollution and offer optimism in the battle against climate change:

- Bioremediation involves employing microorganisms to remediate environmental pollutants, including oil spills, heavy metals, and industrial waste. Some microorganisms possess the ability to break down or render harmless toxic materials, transforming pollutants into safer compounds.

- For instance, following significant oil spills, such as the 2010 Deepwater Horizon disaster, microorganisms like Alcanivorax were crucial in degrading oil in oceanic ecosystems. These microorganisms utilise hydrocarbons, the primary element of oil, transforming them into benign substances such as carbon dioxide and water.

- Carbon sequestration via microbial processes provides a method for capturing and storing carbon. Microorganisms found in soil and ocean sediments have the ability to sequester carbon within organic matter or transform CO_2 into stable, long-lasting carbon reservoirs, contributing to the mitigation of atmospheric concentrations of this greenhouse gas.

Aquatic Microbial Studies

Marine microbiology involves examining the tiny life forms that inhabit oceanic ecosystems, where they are crucial for nutrient cycling, oxygen generation, and sustaining the marine food web. Microorganisms in the ocean play a crucial role in maintaining the health of marine ecosystems and have a significant impact on the entire planet.

Microorganisms in Marine Environments: Their Importance in Nutrient Cycles and Aquatic Food Networks

Marine microbes play a crucial role in oceanic nutrient cycles, making sure that vital elements such as nitrogen, carbon, and phosphorus are accessible to support marine ecosystems.

- Phytoplankton, minuscule photosynthetic entities, constitute the foundation of the marine food web. In the process of photosynthesis, phytoplankton transform carbon dioxide and sunlight into organic matter, providing nourishment for larger organisms such as zooplankton, fish, and marine mammals.

- Marine microorganisms play a crucial role in essential nutrient cycles, including processes like nitrogen fixation, nitrification, and denitrification. Some bacteria, such as Trichodesmium, have the remarkable ability to transform atmospheric nitrogen ($N2$) into a usable form for marine life, enabling the synthesis of proteins and DNA.

- Decomposers in the ocean play a crucial role in breaking down dead organisms and organic material, effectively recycling nutrients back into the ecosystem. In marine ecosystems, bacteria and fungi play a crucial role in breaking down waste products, as well as dead plants and animals. This decomposition process releases essential nutrients that are subsequently absorbed by phytoplankton and other primary producers.

The microbial food web represents a complex network where microorganisms play a crucial role in sustaining the entire marine ecosystem. Without these tiny organisms, the intricate balance of marine ecosystems would unravel, leading to declines in fish populations and jeopardising human food security.

The Importance of Phytoplankton in Oxygen Generation and as the Foundation of the Marine Food Web

One of the most essential functions of marine microorganisms, especially those that photosynthesise, is their role in generating oxygen. Phytoplankton are responsible for approximately fifty percent of global photosynthesis, generating oxygen as a byproduct while also capturing carbon dioxide during this vital process.

- Oxygen production: Phytoplankton are responsible for generating around 50-70% of the Earth's oxygen supply, which is essential for sustaining life on our planet. The process of oxygen production takes place as phytoplankton harness sunlight and carbon dioxide to generate energy, subsequently releasing oxygen into both the atmosphere and the ocean.

- Foundation of the oceanic food web: Phytoplankton, as the foundational producers in oceanic environments, serve as the essential energy source for marine ecosystems. Zooplankton consume phytoplankton, and subsequently, they become prey for larger creatures such as fish, seabirds, and marine mammals. The vitality of global fisheries, which sustain millions of people around the world, relies heavily on the productivity of phytoplankton.

Phytoplankton contribute to climate regulation by affecting the ocean's capacity to absorb CO_2. Through the biological process, phytoplankton capture carbon in organic material, which eventually sinks to the ocean floor, removing CO_2 from the atmosphere and storing it in ocean sediments for long periods.

Microorganisms play an essential role in sustaining environmental health, contributing to nutrient cycles, reducing pollution, and supporting marine ecosystems. Their actions influence the climate, support ecosystems in oceans and forests, and offer answers to critical environmental issues, including climate change and pollution. Grasping and utilising the potential of microorganisms will be essential for fostering a sustainable future for our planet.

CHAPTER 8

Microbiology and Future Technologies

The convergence of microscopic life studies and technological innovation is a swiftly advancing domain that offers the potential for groundbreaking progress across multiple industries, such as healthcare, farming, and energy generation. Through the manipulation of microorganisms, researchers can create groundbreaking solutions to some of the most urgent issues facing our planet. This chapter delves into the fascinating realms of synthetic biology and genetic engineering, the advancement of microbial fuel cells and renewable energy, and the crucial contributions of microbes to space exploration.

Synthetic Biology and Genetic Modification

Synthetic biology represents a convergence of biological sciences and engineering, focussing on the creation and assembly of novel biological components, tools, and systems. This technology allows researchers to alter microbial life at the genetic level, leading to a range of applications that can significantly influence various fields.

How Researchers Alter Microorganisms for Diverse Uses (e.g., CRISPR Technology)

A major advancement in the field of genetic manipulation is the emergence of CRISPR-Cas9 technology. This groundbreaking tool enables researchers to accurately modify the genetic material of a range of organisms, such as bacteria, yeast, and plants.

- CRISPR-Cas9 operates by utilising a natural defence strategy observed in bacteria, which employ RNA molecules to identify and cleave viral DNA. By programming this system to target specific DNA sequences, researchers can add, remove, or modify genes within the genome of a microorganism, resulting in improved traits or functions.

- In microbiology, CRISPR technology has been utilised to modify bacteria for the production of valuable compounds, including pharmaceuticals, biofuels, and biodegradable plastics. For example, scientists have altered Escherichia coli to enhance the production of the anti-cancer drug paclitaxel, providing a more sustainable option compared to conventional production techniques.

- Furthermore, synthetic biology enables the development of microorganisms featuring unique metabolic pathways. Through the integration of synthetic genes, researchers can engineer microorganisms capable of transforming waste products into valuable materials, like converting carbon dioxide into biofuels or synthesising pharmaceuticals from renewable resources.

Prospective Developments in Healthcare, Farming, and Energy Generation

The possibilities for utilising synthetic biology and genetic engineering in fields such as medicine, agriculture, and energy are extensive and revolutionary.

- In biology, engineered microbes could lead to breakthroughs in disease treatment and prevention. For example, researchers are creating probiotics that have been genetically altered to generate therapeutic proteins, aimed at specific conditions such as diabetes or inflammatory bowel disease. These probiotics can provide targeted treatment to the gut, presenting an innovative strategy for addressing chronic conditions.

- In the realm of agriculture, the application of synthetic biology offers the potential to create crops that exhibit greater resilience to environmental challenges, including drought and pest pressures. Through the manipulation of microorganisms that support plant development or improve nutrient absorption, researchers can establish eco-friendly agricultural methods that lessen reliance on synthetic fertilisers and pesticides.

- In the field of energy production, genetically altered microorganisms can act as effective biofuel generators. Investigations are in progress to engineer microorganisms that can efficiently transform sunlight and carbon dioxide into biofuels, presenting a sustainable energy alternative that may diminish dependence on fossil fuels.

With the ongoing progress in synthetic biology, the potential to utilise microbial engineering for tackling global issues will grow, leading to a more sustainable and healthier future.

Microbial Fuel Cells and Sustainable Energy

Microbial fuel cells (MFCs) represent a groundbreaking approach that harnesses the power of microorganisms to transform organic matter into electrical energy. By utilising the metabolic activities of microorganisms, researchers can create sustainable energy solutions that aid in the generation of renewable energy.

Utilising Microbial Potential for Electricity and Biofuel Production

Microbial fuel cells function by harnessing the metabolic activities of bacteria to oxidise organic substrates, which results in the release of electrons. These electrons can be harnessed to produce electricity.

- In a microbial fuel cell, microorganisms are introduced into the anode chamber, where they break down organic materials like wastewater or biomass through metabolic processes. Throughout this process, the microorganisms convey electrons to the anode, resulting in the production of an electric current.

- This technology shows great potential for the treatment of wastewater. Utilising organic pollutants in wastewater as fuel, MFCs generate electricity while simultaneously aiding in water purification, fulfilling a dual role of energy production and pollution mitigation.

- Beyond the generation of electricity, specific microorganisms can be modified to create biofuels. For instance, certain types of algae can be grown to generate biodiesel via photosynthesis, and particular strains of bacteria can be modified to transform sugars into ethanol or butanol, providing sustainable options to replace fossil fuels.

Exploring the Role of Microorganisms in Developing Sustainable Energy Solutions

The possibilities presented by algae and bacteria as sustainable energy sources are vast. Algae have garnered considerable interest because of their rapid growth and substantial lipid content, which can be transformed into biofuels.

- Cultivating algae can occur on land that isn't suitable for traditional agriculture and demands significantly less freshwater than conventional crops, positioning it as a sustainable alternative for biofuel production. Additionally, algae have the ability to take in carbon dioxide during the process of photosynthesis, which plays a role in reducing greenhouse gases.

- Microorganisms also hold promise for sustainable energy production. For instance, specific strains of Clostridium and Escherichia coli can be modified to generate biofuels directly from agricultural waste or even from carbon dioxide. Through the optimisation of microbial pathways, scientists seek to improve the efficiency of biofuel production processes.

The incorporation of microbial fuel cells into waste treatment systems and biofuel production has the potential to greatly diminish our carbon footprint and foster a more sustainable energy environment. As investigations advance, microbial technologies could emerge as essential elements of the worldwide energy landscape.

The Importance of Microorganisms in Space Missions

The investigation of outer space presents distinct challenges, especially in relation to life support systems and the sustainability of extended missions. Microbes can significantly contribute to addressing these challenges, offering solutions for food production, waste recycling, and life support systems.

Exploring Microbial Existence Beyond Earth

The exploration of microbial organisms beyond our planet has seen remarkable growth in recent years. Comprehending the behaviour of microorganisms in microgravity is crucial for the success of future extended space missions.

- Research conducted on the International Space Station (ISS) has demonstrated that certain microorganisms can flourish in microgravity, displaying modified growth behaviours and metabolic processes. Examining these adaptations can reveal important information about microbial resilience and the possibilities of utilising microbes in environments beyond our planet.

- Studying how microorganisms behave in space enhances our knowledge in the quest for extraterrestrial life. Investigating extremophiles—microorganisms that flourish in extreme Earth conditions—allows researchers to formulate theories regarding the possibility of life existing on other planets with severe environments, like Mars or Europa.

Exploring the Role of Microorganisms in Extended Space Expeditions (Bioreactors, Waste Management)

Microbes may play a crucial role in sustaining human life during extended space missions, especially within bioreactors and waste recycling systems.

- Bioreactors employing microorganisms have the potential to supply sustenance and oxygen for astronauts. For example, algae can be grown in bioreactors to generate oxygen via photosynthesis while also acting as a food source. In a similar vein, modified microorganisms have the potential to transform waste materials into valuable resources, thereby establishing a self-sustaining life support system.

- Waste recycling systems utilising microorganisms have the potential to effectively decompose organic waste generated by astronauts, transforming it into valuable nutrients or biofuels. This approach would significantly reduce waste and enhance the efficiency of resource utilisation, thereby ensuring that extended space missions are more sustainable and less reliant on resupply missions from Earth.

- Furthermore, microorganisms could assist in in-situ resource utilisation (ISRU), where local resources are harnessed to support human activities in space. For instance, microorganisms might have the capability to derive nutrients from the soil on Mars or convert carbon dioxide present in the Martian atmosphere into food or fuel.

The incorporation of microbiological principles into new technologies presents extraordinary opportunities for tackling worldwide issues. The advancement of medical therapies, improvement of agricultural practices, provision of renewable energy solutions, and facilitation of space exploration all hinge on the manipulation and understanding of microorganisms, which will play a vital role in crafting a sustainable and technologically sophisticated future. As exploration in these domains advances, we are poised to experience a groundbreaking phase of creativity fuelled by the capabilities of microorganisms.

CHAPTER 9

Getting Started in Microbiology

The study of microorganisms is a multifaceted and swiftly advancing discipline that has gained significance in numerous areas of science and everyday existence. This chapter offers a fundamental insight into the key techniques and tools utilised in the study of microorganisms, catering to those who are new to the field, students, or anyone with a keen interest in the microscopic realm. Additionally, we will delve into new areas and the future of microbiology, emphasising its importance in enhancing our comprehension of life and promoting human well-being.

Fundamental Laboratory Methods

To embark on your exploration in microbiology, it is essential to gain proficiency in basic laboratory techniques. These methods enable scientists to examine and analyse microorganisms while also guaranteeing the secure management of potentially dangerous pathogens.

Overview of Microscopy, Staining Methods, and Cultivation of Microorganisms

Microscopy serves as a fundamental instrument in the study of microscopic life, allowing researchers to observe tiny organisms that cannot be seen without aid. Different kinds of microscopes are employed in the study of microorganisms:

- Optical Instruments: These instruments are widely used and can enlarge specimens by a factor of 1,000. They utilise visible light along with a combination of lenses to enhance the clarity of images. Fundamental light microscopy is crucial for examining the shapes, sizes, and arrangements of bacteria.

- Electron Microscopes: To achieve higher magnification levels, reaching up to 1,000,000 times, electron microscopes are employed. They employ electron beams instead of light, enabling the observation of complex cellular architectures.

Staining techniques play a vital role in microbiology, significantly improving contrast and visibility when observed under the microscope. Typical techniques for staining consist of:

- Gramme Staining: This differential staining method classifies bacteria into two groups, Gram-positive and Gram-negative, depending on the structure of their cell walls. Gram-positive bacteria hold onto the crystal violet stain, resulting in a purple appearance, whereas Gram-negative bacteria do not retain the stain and take on a pink hue after being counterstained with safranin.

- Acid-Fast Staining: This technique is employed to detect mycobacteria, including Mycobacterium tuberculosis. Acid-fast bacteria maintain the red stain (carbolfuchsin) even after being subjected to decolorisation with acid-alcohol, setting them apart from other bacterial types.

Cultivating microorganisms entails nurturing them in regulated environments to examine their traits. This can be accomplished through various forms of media:

- Agar Plates: Solid media that supply essential nutrients for the growth of microorganisms, facilitating the isolation of pure cultures.

- Broth Cultures: Liquid media that facilitate the growth of microorganisms in suspension, providing valuable insights into their metabolic activities.

- Specialised Growth Media: These media promote the proliferation of certain microorganisms while suppressing others, facilitating the examination of specific species.

Maintaining proper aseptic techniques is crucial to avoid contamination while culturing. This involves the careful sterilisation of tools and growth media, as well as operating in proximity to a flame to generate an updraft that reduces the presence of airborne contaminants.

Ensuring Safety in Microbiology Laboratories: Managing Pathogens and Upholding Sterile Conditions

Ensuring safety is crucial in laboratories focused on microorganisms, especially when handling pathogenic strains. Scientists must follow rigorous safety guidelines to ensure their own safety and that of others:

- Safety Levels in Biological Research: Laboratories are categorised into different biosafety levels (BSL 1-4) according to the types of pathogens they manage. BSL-1 laboratories handle agents that are well-understood and present minimal risk, whereas BSL-4 laboratories engage with high-risk pathogens capable of causing serious diseases.

- Safety Gear: Protective gear such as lab coats, gloves, and face shields is crucial for safeguarding researchers against harmful microorganisms.

- Methods of Sterilisation: Autoclaving serves as the conventional approach for ensuring the sterility of equipment and media. It employs high-pressure steam to eliminate all types of microbial organisms. Furthermore, chemical disinfectants play a crucial role in the sanitation of surfaces and equipment.

Ensuring a sterile environment is essential for the success of microbiological studies. This entails consistent maintenance of work areas, appropriate handling of hazardous biological waste, and confirming that all materials are sanitised prior to utilisation.

Instruments for Investigating Microbial Life

Different tools are crucial for examining microorganisms, with each designed to fulfil particular roles that improve research effectiveness.

Examination of Essential Tools Such as Microscopes, Autoclaves, and Incubators

1. Optical Instruments: As previously mentioned, microscopes are essential tools in the study of microorganisms. Light and electron microscopes offer varying degrees of magnification and detail, allowing scientists to examine cellular structures and morphology closely.

2. Autoclaves: These instruments play a vital role in the sterilisation of tools and growth media, guaranteeing the complete removal of any microbial contaminants prior to their application.

3. Incubation Devices: Incubators are essential for creating ideal growth environments for microorganisms, ensuring that temperatures and humidity levels are precisely controlled. These are crucial for cultivating microorganisms, facilitating the examination of their growth dynamics and metabolic processes.

4. Centrifuges: These tools segregate elements of a mixture according to their density. Centrifuges play a crucial role in microbiology, serving to isolate cellular components and purify DNA and proteins effectively.

5. PCR Machines (Thermal Cyclers): These devices amplify specific DNA sequences, facilitating the identification and examination of microorganisms at the genetic level.

Methods for Separating and Recognising Various Microbial Forms

The processes of isolating and identifying microorganisms are essential components of the study of microscopic life forms. A range of methods are utilised for this objective:

- Streak Plate Technique: This method entails spreading a diluted microbial sample over the surface of an agar plate to achieve the isolation of distinct colonies. Every colony originates from a solitary cell, facilitating the establishment of a pure culture.

- Progressive Dilution: This technique entails the dilution of a microbial sample across multiple tubes to lower the concentration of microorganisms, facilitating their isolation in culture.

- Cellular Methods: Advanced techniques, including PCR and DNA sequencing, are becoming more prevalent for identifying microorganisms. These methods enable accurate identification and detailed analysis of microorganisms through their genetic information.

- Biochemical Analyses: Different biochemical tests, such as fermentation tests and enzyme assays, can assist in identifying the unique metabolic capabilities of microorganisms, which is essential for their classification.

The Future of Microbial Science

The domain of microbiology is perpetually advancing, propelled by technological innovations and an increasing comprehension of microbial existence. Numerous new disciplines are transforming the realm of microbiology, presenting thrilling opportunities ahead.

Innovative Areas: Metagenomics, Microbial Ecology, and Tailored Healthcare

1. Metagenomics: This discipline encompasses the examination of genetic material obtained directly from environmental samples. Metagenomic methods enable scientists to examine the complete microbial community found in a sample, offering valuable insights into microbial diversity and functionality. This holds considerable importance for comprehending intricate ecosystems, human microbiomes, and the interactions among microorganisms.

2. Microbial Ecology: This developing field examines the relationships between tiny organisms and the ecosystems they inhabit. Microbial ecologists investigate the role of microbes in nutrient cycling, ecosystem dynamics, and the effects of environmental changes on microbial communities. Grasping these interactions is

essential for tackling environmental issues and ensuring the sustainable management of ecosystems.

3. Tailored Healthcare Solutions: The incorporation of microbiological principles into medical practice is leading to tailored treatment strategies. The study of the human microbiome has uncovered its significance in health and disease, paving the way for possible uses in diagnostics, therapeutics, and disease prevention. For example, analysing the microbiome can inform tailored dietary suggestions or probiotic treatments to enhance health results.

The Ongoing Influence of Microbiology on Our Comprehension of Life and Science

The future of microbiology holds great promise, with its ability to influence a variety of domains, such as healthcare, farming, ecological studies, and biotechnological advancements. Continuous exploration and innovative technologies are set to enhance our comprehension of microbial existence and its influence on ecosystems and human well-being.

- Health Management and Intervention: Ongoing investigation into the human microbiome could reveal innovative approaches for disease prevention and treatment. For instance, comprehending the role of gut microbiota in shaping immune responses may pave the way for innovative immunotherapies or specific strategies for addressing autoimmune diseases.

- Eco-friendly Approaches: The study of microorganisms is essential for advancing sustainable farming methods, boosting soil vitality, and increasing crop production by leveraging the advantages of helpful microbes. These methods can enhance food availability while reducing ecological consequences.

- Innovations in Biotechnology: Progress in microbial biotechnology will persist in generating groundbreaking solutions for energy generation, waste treatment, and ecological restoration. Modifying microorganisms for targeted applications will play a significant role in advancing sustainable technologies that tackle worldwide issues.

Embarking on a journey in microbiology necessitates the acquisition of fundamental laboratory skills, a comprehension of the instruments used for examining microorganisms, and an awareness of the emerging trends within the discipline. As we explore the intricate details of the microscopic realm, the possibilities for this field to enhance our comprehension of life and its uses in diverse areas are expanding significantly. With continuous exploration and innovation, the future of this field holds the potential to reveal new opportunities that will serve both humanity and the Earth at large.

THE END

Made in the USA
Las Vegas, NV
04 January 2025

15881056R00049